FAST MENTAL CALCULATION TRICKS

EasyMath

© 2020 EasyMath. All rights reserved.

I edition, December 2020

ISBN 9798582270102

Contents

Contents	. . . .	5
Introduction	. . . .	6
1	Preliminary remarks	9
2	Some examples	23
3	Add 10, 100, 1000	25
4	Subtract 10, 100, 1000 . . .	29
5	Multiply by 10, 100, 1000 . .	33
6	Divide by 10, 100, 1000 . . .	37
7	Add two numbers	41
8	Subtract two numbers	49
9	Double of a number	57
10	Half of a number	61
11	Quadruple a number and associated calculations	65

1) Preliminary remarks

Before starting, let's review some basic notions, such as multiplication and addition times tables.

First of all we must remember the result of adding two single-digit numbers, ranging from 0 to 9.

Preliminary remarks

Here is the addition table:

and below is the multiplication table:

It is also necessary to remember the power of 2, which we summarize below:

2^0	2^1	2^2	2^3	2^4	2^5	2^6	2^7	2^8	2^9	2^{10}
1	2	4	8	16	32	64	128	256	512	1024

the power of 3:

3^0	3^1	3^2	3^3	3^4	3^5	3^6	3^7	3^8
1	3	9	27	81	243	64	128	729

the squares of the first natural numbers:

1^2	2^2	3^2	4^2	5^2	6^2	7^2	8^2	9^2
1	4	9	16	25	36	49	64	81

the cubes of the first natural numbers:

1^3	2^3	3^3	4^3	5^3	6^3	7^3	8^3	9^3
1	8	27	64	125	216	343	512	729

and the main prime numbers:

$$2, 3, 5, 7, 11, 13, 17, 19, 23, 29, 31, 37, 41,$$
$$43, 47, 53, 59, 61, 67, 71, 73, 79, 83, 89, 97.$$

The properties of the basic operations, i.e., the commutative property, the associative property and the distributive property, will also be useful for learning fast calculation tricks.

The commutative property concerns addition and multiplication. Here we propose

Preliminary remarks

some examples:

$$5 + 3 = 3 + 5,$$
$$2 + 1 = 1 + 2,$$

where we see that by exchanging the order of the addends the result does not change and

$$5 \times 3 = 3 \times 5,$$
$$2 \times 1 = 1 \times 2,$$

i.e., exchanging the order of the factors does not change the result.

The associative property concerns only addition and multiplication. For example:

$$5 + 3 + 2 = (5 + 3) + 2 = 5 + (3 + 2),$$
$$2 + 1 + 7 = (2 + 1) + 7 = 2 + (1 + 7),$$

we can therefore firstly add some addends together and then add the result to the remaining addend without changing the result of the calculation and, in a similar way

$$5 \times 3 \times 2 = (5 \times 3) \times 2 = 5 \times (3 \times 2),$$
$$2 \times 1 \times 7 = (2 \times 1) \times 7 = 2 \times (1 \times 7),$$

i.e., in a multiplication of three or more terms we can replace some terms with their product, obtaining the same result.

This property can also be used in the reverse sense, that is, for example

$$8 + 2 = (5 + 3) + 2 = 5 + 3 + 2,$$
$$3 + 7 = (2 + 1) + (5 + 2) = 2 + 1 + 5 + 2,$$

or, for multiplication,

$$15 \times 2 = (5 \times 3) + 2 = 5 \times 3 \times 2,$$

in this case we can call it dissociative property.

The third property, the distributive one, refers to both addition and multiplication operations, at the same time. Let's see some examples

$$2 \times (3 + 5) = 2 \times 3 + 2 \times 5,$$
$$3 \times (1 + 6 + 4) = 3 \times 1 + 3 \times 6 + 3 \times 4$$

and

$$7 \times (1 + 2 + 6 + 9) = 7 \times 1 + 7 \times 2$$
$$+ 7 \times 6 + 7 \times 9,$$

or

$$2 \times 1 + 2 \times 2 + 2 \times 4 = 2 \times (1 + 2 + 4).$$

This property therefore states that when multiplying and adding together several numbers (using brackets, to indicate the order in which the operations are carried out) we can multiply the same number for each of the addends and add the various partial products obtained.

In fast mental calculation we are obviously dealing with numbers. A number can be integer (without a comma), such as 22 or a decimal which in turn is divided into an integer part, placed before the comma and a decimal part, after the comma, such as 33,7. The digits of a number have a value that depends on the position occupied in the number itself, called the position value of the digit.

In the number 372,146 the integer part is 372, while the decimal part is 146. Furthermore, the 6 digits occupy positions to which a name has also been assigned to remember them better. The 3 represents the hundreds, the 7 the tens, the 2 the ones, the 1 the tenths, the 4 the hundredths and the 6 the thousandths.

We remember that

- 1 thousand = 1000;

- 1 hundred = 100;

- 1 ten = 10;

- 1 one = 1;

- 1 tenth = 0,1 = 1/10;

- 1 hundredth = 0,01 = 1/100;

- 1 thousandth = 0,001 = 1/1000.

In a number, moving from left to right the position value of the digits decreases ten times (a factor 1/10) for each place, while moving from right to left the position value increases ten times (a factor 10) for each place.
To move from one place to the next you need a quantity of "10" units of the previous place, for example:

10 tens = 1 hundred ($10 \times 10 = 100$);
10 ones = 1 ten ($10 \times 1 = 10$);
100 thousandths = 1 tenth ($100 \times 0,001 = 0,1$).

A number can also be written in a form, called polynomial form (it recalls the concept of polynomial), where ones, tens, hundreds, thousands, and so on are added together. For example, the number 274 can be written as

$$274 = 2 \times 100 + 7 \times 10 + 4 = 200 + 7 + 4\,,$$

in fact it is made up of 2 hundreds, 7 tens and 4 ones.

This form is very important because in many cases it can really simplify the calculations. A few more examples to better fix the ideas:

1. Number: 8253:

- 8 thousands, 2 hundreds, 5 tens, 3 ones;

- polynomial form:

$$8 \times 1000 + 2 \times 100 + 5 \times 10 + 3$$
$$= 8000 + 200 + 50 + 3\,;$$

2. Number: 372:

   - 3 hundreds, 7 tens, 2 ones;

   - polynomial form:

   $$3 \times 100 + 7 \times 10 + 2 = 300 + 70 + 2\,;$$

3. Number: 78,989:

   - 7 tens, 8 ones, 9 tenths, 8 cents, 9 thousandths;

- polynomial form:

$$7 \times 10 + 8 + 9 \times 0{,}1$$
$$+ 8 \times 0{,}01 + 9 \times 0{,}001$$
$$= 70 + 8 + 0{,}9 + 0{,}08 + 0{,}009 \,.$$

2) Some examples

Thanks to what you will read later you will know immediately how to solve problems such as:

- Which of the following answers represents the correct result for the product 526 × 812?
  a) 419374;
  b) 427112;
  c) 430813;
  d) 434365.

It might seem difficult, but the answer is very easy, it is b) 427112, and you will be able to perform the calculation mentally, without the aid of tools, and in a very short time. These are calculations that you were doing with the help of paper and pen or a calculator.

For example, you will be able to make calculations such as

$$230 : 5, \quad 320 \times 11,$$
$$125^2, \quad 21 \times 35,$$

and so on.

3) Add 10, 100, 1000

The first thing to learn is to add multiples of 10, i.e.

$$10, 100, 1000, 10000$$

and so on, to any number.

The procedure is very simple, just take the number in question and increase by 1 the digit that occupies the position occupied by the number 1 in the multiple of 10 considered.

Add 10, 100, 1000

Let's take an example, consider the sum

$$653 + 10.$$

The 1 in 10 occupies the tens, so we need to increase the tens digit of the number 653 by 1, which is 5. We have

$$5 + 1 = 6$$

and therefore, simply,

$$653 + 10 = 663.$$

If the number under consideration is 9, it must become 0 and increase the leftmost digit by 1. To understand better, let's consider the following sum

$$693 + 10.$$

Add 10, 100, 1000

The digit to increase would be that of the tens of 693, which is a 9. We therefore put a 0 in place of 9 and increase the leftmost digit by one, that is, 6, that of the hundreds, which becomes 7. We have

$$693 + 10 = 703.$$

Let's take other examples to understand the particular situations that can happen:

- $2274 + 100 = 2374$ (increase by 1 the digit of the hundreds of 2274, which passes from 2 to 3);

- $873845 + 10000 = 874845$ (increase by 1 the digit 3, to the fourth position from the right, which passes to 4)

- $499 + 10 = 509$ (increase the leftmost digit of the tens by 1, being 9, which goes from

4 to 5 and I put 0 in place of 9 of the tens);

- $987 + 100 = 1087$ (see 987 as 0987 and apply the same rule, i.e. I should increase the hundreds digit by 1, but being a 9, I put a 0 in its place and increase the digit to its left by 1, which being a 0 goes to 1).

If a number is decimal, the reasoning made so far is valid as it is for the integer part, while the decimal part remains unchanged.

4) Subtract 10, 100, 1000

The second calculation to learn is closely related to the one introduced before, i.e. subtract multiples of 10

$$10, 100, 1000, 10000$$

and so on, to any number.

To do this, simply take the number in question and decrease by 1 the digit that occupies the position occupied by the number 1 in the multiple of 10 considered.

Let's take an example, consider the subtrac-

tion
$$761 - 10\,.$$

The 1 in 10 occupies the tens, so we must decrease the tens digit of the number 761 by 1, which is 5. We have

$$6 - 1 = 5$$

and therefore, simply,

$$761 - 10 = 751\,.$$

If the number taken into consideration is 0, it must become 9 and decrease the leftmost digit by 1. To understand better let's consider the following subtraction

$$804 - 10\,.$$

Subtract 10, 100, 1000

The digit to decrease would be that of the tens of 804, which is a 0. We therefore put a 9 in place of the 0 and decrease the leftmost digit by one, that is, the 8, that of the hundreds, which becomes 7. We have

$$804 - 10 = 794.$$

Let's take other examples to understand the various situations that can happen:

- $2274 - 100 = 2174$ (decrease by 1 the digit of the hundreds of 2274, which goes from 2 to 1);

- $873845 - 10000 = 872845$ (decrease by 1 the digit 3, to the fourth position from the right, which passes to 2);

- $409 - 10 = 399$ (decrease by 1 the leftmost digit of that of the tens, being 0, which goes from 4 to 3 and I put 9 in place of 0 of the tens);

- $987 - 100 = 887$ (decrease of 1 the digit of the hundreds of 987, which goes from 9 to 8).

5) Multiply by 10, 100, 1000

Let's start with something very easy, multiplying by 10, by 100, by 1000 and by multiples, in general, of 10.

How to multiply a number by 100? The answer is immediate. If the number is integer just add 2 zeros to the bottom of the number itself, 2 zeros because they are those present in the number 100, for example:

$$276 \times 100 = 27600\,.$$

If, on the other hand, the number is decimal, just move the comma to the right (the value of the number must grow) by 2 places (2 because 100 has 2 zeros), for example:

$$30,44 \times 100 = 3044,$$
$$19,1 \times 100 = 1910,$$

notice how we added a zero at the end, in fact the number 19,1 can be seen as 19,10, or as 19,100, or 19,1000, that are all identical representations of the same number.

Let's generalize this rule. The result of multiplying an integer by a multiple of 10 with n zeros is obtained by rewriting the starting number and adding n zeros to the bottom right.

Multiply by 10, 100, 1000

Examples:

$$37 \times 1000 = 37000,$$
$$2348 \times 10 = 23480,$$
$$370 \times 100 = 37000,$$
$$29864 \times 1000 = 29864000,$$
$$11 \times 10000 = 110000.$$

The result of multiplying a decimal number by a multiple of 10 with n zeros is obtained by rewriting the starting number and moving the comma to the right by n places, adding zeros if necessary.

Multiply by 10, 100, 1000

Examples:

$$3{,}7 \times 1000 = 3700\,,$$
$$2{,}348 \times 10 = 23{,}48\,,$$
$$3{,}71 \times 100 = 371\,,$$
$$298{,}64 \times 1000 = 298640\,,$$
$$1{,}1 \times 10000 = 11000\,.$$

6) Divide by 10, 100, 1000

Let's now pass to the division by 10, by 100, by 1000 and by multiples, in general, of 10. How to divide a number by 10? If the number is an integer just add a comma (and thus the number becomes decimal) just before the ones, for example:

$$276 : 10 = 27{,}6 \,,$$
$$320 : 10 = 32{,}0 = 32 \,,$$

we see that if we divide by 10 a number that ends with 0, we just remove the 0. On the other hand, if the number is decimal, just move the comma to the left (the value of the number must decrease) by 1 place (1 because 10 has 1 zero), for example:

$$30{,}44 : 10 = 3{,}044\,,$$
$$19{,}1 : 10 = 1{,}91\,,$$
$$5{,}287 : 10 = 0{,}5287\,,$$

observe that we added a zero at the beginning, in fact the number 5,287 can be seen as 05,287, or as 005,287, or even 0005,287, that are all identical representations of the same number, not used for obvious reasons of practicality.

Let's generalize the rule. The result of dividing an integer by a multiple of 10 with n zeros is obtained by rewriting the starting number and adding a comma to n places to

Divide by 10, 100, 1000

the left starting from the bottom position, furthest to the right.

Examples:

- 37 : 1000 = 0,037 (comma placed after 3 places from the bottom right);

- 2348 : 10 = 234,8 (comma placed after 1 place from the bottom right);

- 370 : 100 = 3,7 (comma placed after 2 places from the bottom on the right, eliminating the superfluous zero, in fact 3,70 = 3,7);

- 29864 : 1000 = 29,864 (comma placed after 3 places from the bottom right);

- 11 : 10000 = 0,0011 (comma placed after 4 places from the bottom to the right, adding appropriate zeros).

Divide by 10, 100, 1000

The result of dividing a decimal number by a multiple of 10 with n zeros is obtained by rewriting the starting number and moving the comma to the left by n places, adding zeros if necessary.

Examples:

$$9993{,}7 : 1000 = 9{,}9937\,,$$
$$2{,}348 : 10 = 0{,}2348\,,$$
$$3{,}71 : 100 = 0{,}0371\,,$$
$$2987{,}64 : 1000 = 2{,}98764\,,$$
$$1{,}1 : 10000 = 0{,}00011\,.$$

7) Add two numbers

Let's see some tricks to quickly add two numbers. The trick is to break down the smallest addend into tens, hundreds, thousands, and so on.

We propose some examples to better understand how to proceed. Suppose we want to add the number 27 to the number 11, i.e. we want to calculate

$$27 + 11 = ?\,.$$

The smaller of the two is 11, let's separate it into tens and units, we have

$$11 = 10 + 1,$$

that is a ten plus one. So we can write

$$27 + 11 = 27 + 10 + 1.$$

Now let's add 27 and 10, increasing the ten of the 27 by one unit (i.e. the 2 which becomes 3) from which

$$27 + 10 = 37$$

and simply add 1

$$27 + 10 + 1 = 37 + 1 = 38,$$

hence the result

$$27 + 11 = 38\,.$$

Another example:

$$46 + 35 = 46 + 30 + 5$$
$$= 46 + 10 + 10 + 10 + 5 = 56 + 10 + 10 + 5$$
$$= 66 + 10 + 5 = 76 + 5 = 81\,,$$

or again

$$217 + 23 = 217 + 20 + 3 = 237 + 3 = 240\,,$$
$$136 + 47 = 136 + 40 + 7 = 176 + 7 = 183\,,$$

where to calculate

$$176 + 7$$

we added 7 to the ones of 176, i.e.

$$7 + 6 = 13,$$

from which

$$176 + 7 = 170 + 6 + 7 = 170 + 13 = 183.$$

Decomposing a number as the sum of numbers ending with some zero simplifies the calculations considerably.

Still examples:

$$473 + 226 = 473 + 200 + 20 + 6$$
$$= 473 + 100 + 100 + 10 + 10 + 6$$
$$= 673 + 10 + 10 + 6 = 693 + 6 = 699$$

and

$$619 + 314 = 619 + 300 + 10 + 4$$
$$= 619 + 100 + 100 + 100 + 10 + 4$$
$$= 919 + 10 + 4 = 929 + 4 = 933.$$

To add two numbers, one of which is close to 10, 100 or 1000, we can use some easy-to-learn tricks.
Consider for example the sum

$$38 + 12,$$

we can write

$$38 + 10 + 2 = 48 + 2 = 50.$$

Let us now consider the sum

$$67 + 19$$

you can see 19 how

$$19 = 20 - 1$$

and write

$$67 + 20 - 1 = 87 - 1 = 86,$$

in fact adding 67 and 20 is extremely simple and fast, thanks to our base-10 number system.

Other examples involving the various methods exposed:

$$48 + 17 = 48 + 10 + 7 = 58 + 7 = 65,$$
$$48 + 17 = 48 + 20 - 3 = 68 - 3 = 65,$$
$$876 + 102 = 876 + 100 + 2 = 976 + 2 = 878,$$

and

$$876 + 99 = 876 + 100 - 1 = 976 - 1 = 975,$$
$$29 + 18 = 29 + 10 + 8 = 39 + 8 = 47,$$
$$29 + 18 = 29 + 20 - 2 = 49 - 2 = 47.$$

We see that it is preferable to arrive at the nearest multiple of 10 (or 100 or 1000) and then add or subtract a lower number.

These techniques can also be refined for numbers between multiples of 10, as we see in the following examples:

$$58 + 7 = 50 + 8 + 7 = 50 + 5 + 3 + 5 + 2$$
$$= 50 + 10 + 3 + 2 = 60 + 5 = 65,$$

or

$$78 + 9 = 70 + 8 + 5 + 4 = 70 + 5 + 3 + 5 + 4$$
$$= 70 + 10 + 4 + 3 = 80 + 7 = 87,$$

or again,

$$35 + 6 = 30 + 5 + 5 + 1 = 30 + 10 + 1$$
$$= 40 + 1 = 41.$$

8) Subtract two numbers

Let's now deal with the fast subtraction between two numbers. We separate the smallest addend into tens, hundreds, thousands, and so on to facilitate the calculation. Here are some examples. Suppose we want to subtract the numbers 37 and 12, i.e. calculate

$$37 - 12 = ?\,.$$

The smaller of the two is 12, let's separate it into tens and units, i.e.

$$12 = 10 + 2\,,$$

or one ten plus two ones. We can write

$$37 - 12 = 37 - 10 - 2.$$

We then subtract 10 from 37, decreasing the ten of 37 by 1 (i.e. 3 which becomes 2) from which
$$37 - 10 = 27$$

and add 1

$$37 - 10 - 2 = 27 - 2 = 25,$$

hence the result

$$37 - 12 = 25.$$

Other examples:

$$46 - 35 = 46 - 30 - 5$$
$$= 46 - 10 - 10 - 10 - 5$$
$$= 36 - 10 - 10 - 5 = 26 - 10 - 5$$
$$= 16 - 5 = 11,$$

or

$$217 - 23 = 217 - 20 - 3$$
$$= 217 - 10 - 10 - 3$$
$$= 207 - 10 - 3 = 197 - 3 = 194,$$

or

$$136 - 47 = 136 - 40 - 7$$
$$= 136 - 30 - 10 - 7 = 106 - 10 - 7$$
$$= 96 - 7 = 89.$$

Still examples:

$$473 - 226 = 473 - 200 - 20 - 6$$
$$= 473 - 100 - 100 - 10 - 10 - 6$$
$$= 273 - 10 - 10 - 6 = 253 - 6 = 247$$

and

$$619 - 314 = 619 - 300 - 10 - 4$$
$$= 319 - 10 - 4 = 309 - 4 = 305.$$

To subtract two numbers, one of which is close to 10, 100 or 1000, we can use the same methods shown for adding them.

For example, consider subtraction

$$67 - 19,$$

you can see 19 as

$$19 = 20 - 1$$

and write

$$67 - (20 - 1) = 67 - 20 + 1 = 47 + 1 = 48,$$

where we took advantage of the fact that subtracting 20 to 67 is a quick calculation. Other examples:

$$48 - 17 = 48 - 10 - 7 = 38 - 7 = 31,$$
$$876 - 102 = 876 - 100 - 2 = 776 - 2 = 774,$$

or

$$48 - 17 = 48 - (20 - 3) = 48 - 20 + 3$$
$$= 28 + 3 = 31$$

$$876 - 99 = 876 - (100 - 1) = 876 - 100 + 1$$
$$= 776 + 1 = 777,$$

or again

$$29 - 18 = 29 - 10 - 8 = 19 - 8 = 11$$

$$29 - 18 = 29 - (20 - 2) = 29 - 20 + 2$$
$$= 9 + 2 = 11.$$

We see that it is preferable to arrive at the nearest multiple of 10 (or 100 or 1000) and then add or subtract a lower number.

These techniques can also be refined for numbers between multiples of 10, as we see in the following examples:

$$58 + 7 = 50 + 8 + 7 = 50 + 5 + 3 + 5 + 2$$
$$= 50 + 10 + 3 + 2 = 60 + 5 = 65$$

$$78 + 9 = 70 + 8 + 5 + 4 = 70 + 5 + 3 + 5$$
$$+ 4 = 70 + 10 + 4 + 3 = 80 + 7 = 87$$

$$35 + 6 = 30 + 5 + 5 + 1 = 30 + 10 + 1$$
$$= 40 + 1 = 41.$$

9) Double of a number

How do you quickly calculate twice a number (multiplying the number by 2)?
Meanwhile, let's say that the double of a number can be calculated by adding the number to itself.
The double of 13 is obtained by adding up

$$13 \times 2 = 13 + 13 = 26 \,.$$

Let's make some explanatory examples:

$$27 \times 2 = 27 + 27 = 27 + 20 + 7$$
$$= 47 + 7 = 54 \,,$$

Double of a number

$$32 \times 2 = 32 + 32 = 64,$$

$$49 \times 2 = 49 + 49 = 49 + 40 + 9$$
$$= 89 + 9 = 98,$$

$$179 \times 2 = 179 + 179 = 179 + 100 + 79$$
$$= 279 + 70 + 9 = 349 + 9 = 358.$$

Another technique to obtain the double of a number is to calculate double each of its positional digits (double the ones, double the tens, double the hundreds, and so on) and add the results.

Examples:

$$27 \times 2 = (20 + 7) \times 2$$
$$= 20 2 + 7 \times 2 = 40 + 14 = 54,$$

$$32 \times 2 = (30 + 2) \times 2$$
$$= 30·2 + 2 \times 2 = 60 + 4 = 64,$$

$$49 \times 2 = (40 + 9) \times 2$$
$$= 40·2 + 9 \times 2 = 80 + 18 = 98,$$

$$179·2 = (100 + 70 + 9) \times 2$$
$$= 100·2 + 70·2 + 9·2 = 200 + 140 + 18$$
$$= 340 + 18 = 358.$$

It is also possible to perform some of the previous calculations as follows

$$27 \times 2 = (30 - 3) \times 2$$
$$= 30·2 - 3 \times 2 = 60 - 6 = 54,$$

Double of a number

$$49 \times 2 = (50 - 1) \times 2$$
$$= 502 - 1 \times 2 = 100 - 2 = 98,$$

$$1792 = (100 + 80 - 1) \times 2$$
$$= 1002 + 802 - 12 = 200 + 160 - 2$$
$$= 360 - 2 = 358.$$

10) Half of a number

Let's now quickly calculate the half of a number (dividing that number by 2).

First of all we claim that the half of an even number is an integer, while the half of an odd number is a decimal number, with a fixed decimal part equal to 5. For example, the half of 8 is 4 and the half of 7 is 3,5.

The trick for obtain the half of a number is to calculate the half of each of its positional digits (half of the ones, half of the tens, half of the hundreds, and so on) and add the results. Consider the following examples to better understand:

$$27 : 2 = (20 + 7) : 2 = 20 : 2 + 7 : 2$$
$$= 10 + 3{,}5 = 13{,}5\,,$$

$$32 : 2 = (30 + 2) : 2 = 30 : 2 + 2 : 2$$
$$= 15 + 1 = 16\,,$$

$$49 : 2 = (40 + 9) : 2 = 40 : 2 + 9 : 2$$
$$= 20 + 4{,}5 = 24{,}5$$

$$179 : 2 = (100 + 70 + 9) : 2$$
$$= 100 : 2 + 70 : 2 + 9 : 2$$
$$= 50 + 35 + 4{,}5 = 85 + 4{,}5 = 89{,}5\,,$$

or also

$$27 : 2 = (30 - 3) : 2 = 30 : 2 - 3 : 2$$
$$= 15 - 1{,}5 = 13{,}5\,,$$

$$49 : 2 = (50 - 1) : 2 = 50 : 2 - 1 : 2$$
$$= 25 - 0{,}5 = 24{,}5\,,$$

$$179 : 2 = (100 + 80 - 1) : 2$$
$$= 100 : 2 + 80 : 2 - 1 : 2$$
$$= 50 + 40 - 0{,}5 = 90 - 0{,}5 = 89{,}5\,.$$

For convenience, we report the half of the 9 digits in the following table:

0/2	1/2	2/2	3/2	4/2	5/2	6/2	7/2	8/2	9/2
0	0,5	1	1,5	2	2,5	3	3,5	4	4,5

11) Quadruple a number and related calculations

Quadrupling a number is multiplying it by four and the result can be calculated simply by considering it as the sum of its double with the double itself, or as double the double of the number.

Let's see some examples, in pairs, using the two ways to proceed:

Example 1:

Quadruple a number and associated calcula-...

- sum of the doubles:

$$14 \times 4 = (14 \times 2) + (14 \times 2) = 28 + 28$$
$$= 28 + 20 + 8 = 48 + 8 = 56\,;$$

- double the double:

$$14 \times 4 = (14 \times 2) \times 2$$
$$= 28 \times 2 = (20 + 8) \times 2$$
$$= 20 \times 2 + 8 \times 2 = 40 + 16 = 56\,.$$

Example 2:

- sum of doubles:

$$17 \times 4 = (17 \times 2) + (17 \times 2) = 34 + 34$$
$$= 34 + 30 + 4 = 64 + 4 = 68\,;$$

- double the double:

$$17 \times 4 = (17 \times 2) \times 2 = 34 \times 2$$
$$= (30 + 4) \times 2 = 30 \times 2 + 4 \times 2$$
$$= 60 + 8 = 68\,.$$

Example 3:

- sum of doubles:

$$59 \times 4 = (59 \times 2) + (59 \times 2) = 118 + 118$$
$$= 118 + 100 + 10 + 8 = 218 + 10 + 8$$
$$= 228 + 8 = 236\,;$$

Quadruple a number and associated calcula-...

- double the double:

$$59 \times 4 = (59 \times 2) \times 2 = 118 \times 2$$
$$= (100 + 10 + 8) \times 2$$
$$= 100 \times 2 + 10 \times 2 + 8 \times 2$$
$$= 200 + 20 + 16 = 220 + 16 = 236 \,,$$

where we used the calculation

$$59 \times 2 = (60 - 1) \times 2$$
$$= 60 \times 2 - 1 \times 2 = 120 - 2 = 118 \,.$$

If we want to divide a number by 4 we can divide it by 2 (calculating the half of the number) and then again divide it by 2 (half again). In practice it is like calculating half of the half. Examples:

$$64 : 4 = (64 : 2) : 2 = 32 : 2 = 16$$

Quadruple a number and associated calcula-...

$$66 : 4 = (66 : 2) : 2 = 33 : 2 = (30 + 3) : 2$$
$$= 30 : 2 + 3 : 2 = 15 + 1{,}5 = 16{,}5 \,,$$

$$148 : 4 = (148 : 2) : 2 = 74 : 2$$
$$= (70 : 2) + (4 : 2) = 35 + 2 = 37 \,,$$

$$1360 : 4 = (1360 : 2) : 2$$
$$= (1000 : 2 + 300 : 2 + 60 : 2) : 2$$
$$= (500 + 150 + 30) : 2 = 680 : 2$$
$$= (600 : 2) + (80 : 2) = 300 + 40 = 340 \,.$$

Multiplying a number by 8 means multiplying it by 4 and then by 2, that is, considering the double of the quadruple (or, again, the quadruple of the double).

Quadruple a number and associated calcula-...

We calculate

$$21 \times 8 = (21 \times 4) \times 2$$
$$= (21 \times 2 + 21 \times 2) \times 2$$
$$= (42 + 42) \times 2 = 84 \times 2 = (80 + 4) \times 2$$
$$= 80 \times 2 + 4 \times 2 = 160 + 8 = 168.$$

Other examples:

$$33 \times 8 = (33 \times 4) \times 2$$
$$= (33 \times 2 + 33 \times 2) \times 2$$
$$= (66 + 66) \times 2 = (60 + 6 + 60 + 6) \times 2$$
$$= (120 + 12) \times 2 = (100 + 20 + 12) \times 2$$
$$= 100 \times 2 + 20 \times 2 + 12 \times 2$$
$$= 200 + 40 + 24 = 240 + 24 = 264,$$

or

$$52 \times 8 = (52 \times 4) \times 2$$
$$= (52 \times 2 + 52 \times 2) \times 2$$
$$= (104 + 104) \times 2 = 208 \times 2$$
$$= (200 + 8) \times 2 = 200 \times 2 + 8 \times 2$$
$$= 400 + 16 = 416.$$

12) Multiply by 5

When you need to multiply a number by 5, simply calculate its half and then multiply the obtained number by 10, in fact 5 is the half of 10.

Let's see some application examples:

$$14 \times 5 = (14 : 2) \times 10 = 7 \times 10 = 70\,,$$
$$15 \times 5 = (15 : 2) \times 10 = 7,5 \times 10 = 75\,,$$
$$26 \times 5 = (26 : 2) \times 10 = 13 \times 10 = 130$$

and again

$$39 \times 5 = (39 : 2) \times 10 = (30 : 2 + 9 : 2) \times 10$$
$$= (15 + 4,5) \times 10 = 19,5 \times 10 = 195\,,$$

Multiply by 5

$$118 \times 5 = (118 : 2) \times 10$$
$$= (100 : 2 + 18 : 2) \times 10$$
$$= (50 + 9) \times 10 = 59 \times 10 = 590\,,$$

$$238 \times 5 = (238 : 2) \times 10$$
$$= (200 : 2 + 38 : 2) \times 10$$
$$= (100 + 19) \times 10 = 119 \times 10 = 1190\,.$$

13) Divide by 5

If we want to divide a number by 5 it is sufficient to simply calculate its double and then divide the obtained result by 10.

Some examples:

$$14 : 5 = (14 \times 2) : 10 = 28 : 10 = 2{,}8\,,$$
$$15 : 5 = (15 \times 2) : 10 = 30 : 10 = 3\,,$$
$$26 : 5 = (26 \times 2) : 10 = 52 : 10 = 5{,}2\,,$$

or

$$39 : 5 = (39 \times 2) : 10$$
$$= (30 \times 2 + 9 \times 2) : 10$$
$$= (60 + 18) : 10 = 78 : 10 = 7{,}8\,,$$

$$95 : 5 = (95 \times 2) : 10$$
$$= (100 \times 2 - 5 \times 2) : 10$$
$$= (200 - 10) : 10 = 190 : 10 = 19\,.$$

$$118 : 5 = (118 \times 2) : 10$$
$$= (100 \times 2 + 18 \times 2) : 10$$
$$= (200 + 36) : 10 = 236 : 10 = 23{,}6\,,$$

Divide by 5

$$238 : 5 = (238 \times 2) : 10$$
$$= (200 \times 2 + 38 \times 2) : 10$$
$$= (400 + 30 \times 2 + 8 \times 2) : 10$$
$$= (400 + 60 + 16) : 10 = (460 + 16) : 10$$
$$= 476 : 10 = 47{,}6 \,.$$

14) Multiply by 1,5

How do you easily multiply a number by 1,5? The answer is very simple, just observe that multiplying by 1,5 is equivalent to calculate the half of the triple, or it can be obtained by adding its half to the starting number. Some examples that make use of the first technique are:

$$12 \times 1{,}5 = (12 \times 3) : 2 = 36 : 2 = 18,$$

Multiply by 1,5

$$17 \times 1{,}5 = (17 \times 3) : 2$$
$$= (10 \times 3 + 7 \times 3) : 2$$
$$= (30 + 21) : 2 = 51 : 2$$
$$= (50 + 1) : 2 = 25 + 0{,}5 = 25{,}5 \,,$$

$$74 \times 1{,}5 = (74 \times 3) : 2$$
$$= (70 \times 3 + 4 \times 3) : 2$$
$$= (210 + 12) : 2 = 222 : 2 = 111 \,,$$

$$99 \times 1{,}5 = (99 \times 3) : 2$$
$$= (100 \times 3 - 1 \times 3) : 2$$
$$= (300 - 3) : 2 = 150 - 1{,}5 = 148{,}5 \,.$$

The same calculations performed with the second technique:

$$12 \times 1{,}5 = 12 + (12 : 2) = 12 + 6 = 18 \,,$$

Multiply by 1,5

$$17 \times 1{,}5 = 17 + (17:2)$$
$$= 17 + (10:2 + 7:2)$$
$$= 17 + 5 + 3{,}5 = 22 + 3{,}5 = 25{,}5\,,$$

$$74 \times 1{,}5 = 74 + (74:2)$$
$$= 74 + (70:2 + 4:2) - 74 + 35 + 2$$
$$= 70 + 30 + 4 + 5 + 2 = 111\,,$$

$$99 \times 1{,}5 = 99 + (99:2)$$
$$= 99 + (100:2 - 1:2) = 99 + 50 - 0{,}5$$
$$= 100 - 1 + 50 - 0{,}5 = 150 - 1{,}5 = 148{,}5\,.$$

15) Multiply by multiples of 10

Suppose we want to calculate the product of a number and a multiple of 10, for example

$$12 \times 30,$$

where 30 is multiple of 10, in fact

$$30 = 3 \times 10,$$

we can write

$$12 \times 30 = (12 \times 3) \times 10 = 36 \times 10 = 360.$$

Multiply by multiples of 10

We observe that the product between a number and n times 10 can be obtained just calculating the product of the number by n and then multiply the result by 10.

Other explanatory examples:

$$8 \times 50 = (8 \times 5) \times 10 = 40 \times 10 = 400,$$
$$9 \times 90 = (9 \times 9) \times 10 = 81 \times 10 = 810,$$

$$15 \times 40 = (15 \times 4) \times 10$$
$$= (15 \times 2 \times 2) \times 10$$
$$= (30 \times 2) \times 10 = 60 \times 10 = 600,$$

$$81 \times 20 = (81 \times 2) \times 10$$
$$= (80 \times 2 + 1 \times 2) \times 10$$
$$= (160 + 2) \times 10 = 162 \times 10 = 1620,$$

Multiply by multiples of 10

$$24 \times 70 = (24 \times 7) \times 10$$
$$= (20 \times 7 + 4 \times 7) \times 10$$
$$= (2 \times 7 \times 10 + 28) \times 10$$
$$= (14 \times 10 + 28) \times 10$$
$$= (140 + 28) \times 10 = 168 \times 10 = 1680\,.$$

16) Multiply by 11, 21, 31, ⋯

Let's see other examples of fast calculations. To multiply a number by 11 just multiply it by 10 and add the number itself, in fact

$$11 = 10 + 1\,.$$

Let's see some examples:

$$23 \times 11 = (23 \times 10) + 23 = 230 + 23 = 253\,,$$
$$40 \times 11 = (40 \times 10) + 40 = 400 + 40 = 440\,,$$

$$88 \times 11 = (88 \times 10) + 88 = 880 + 88$$
$$= 900 - 20 + 80 + 8 = 980 - 20 + 8$$
$$= 960 + 8 = 968,$$

$$145 \times 11 = (145 \times 10) + 145$$
$$= 1450 + 145 = 1450 + 100 + 40 + 5$$
$$= 1550 + 40 + 5 = 1590 + 5 = 1595.$$

Similarly, to multiply a number by 21 just multiply it by 20 and add the number itself, in fact

$$21 = 20 + 1,$$

where, in turn,

$$20 = 2 \times 10.$$

Multiply by 11, 21, 31, ···

Examples:

$$23 \times 21 = (23 \times 20) + 23$$
$$= (23 \times 2) \times 10 + 23$$
$$= 46 \times 10 + 23 = 460 + 23 = 483,$$

$$56 \times 21 = (56 \times 20) + 56$$
$$= (56 \times 2) \times 10 + 56$$
$$= (50 \times 2 + 6 \times 2) \times 10 + 56$$
$$= (100 + 12) \times 10 + 56 = 1120 + 56$$
$$= 1120 + 50 + 6 = 1170 + 6 = 1176,$$

$$83 \times 21 = (83 \times 20) + 83$$
$$= (83 \times 2) \times 10 + 83$$
$$= (80 \times 2 + 3 \times 2) \times 10 + 83$$
$$= (160 + 6) \times 10 + 83 = 1660 + 83$$
$$= 1660 + 80 + 3 = 1740 + 3 = 1743 \,.$$

The multiplication of a number by 31 (and, similarly, by 41, 51, 61 and so on) can be done considering that

$$31 = 30 + 1 = 3 \times 10 + 1$$

and then multiply the number by 30 and add the result to the number itself. For example

$$7 \times 31 = (7 \times 30) + 7 = 210 + 7 = 217 \,,$$

Multiply by 11, 21, 31, $\cdots$

$$22 \times 31 = (22 \times 30) + 22$$
$$= (22 \times 3) \times 10 + 22$$
$$= 66 \times 10 + 22 = 660 + 22 = 682.$$

17) Multiply by 9, 19, 29, ⋯

Similar calculations to the previous ones are those for 9, 19, 29 and so on.
To multiply a number by 9 just multiply it by 10 and subtract the number itself, in fact

$$9 = 10 - 1.$$

Some examples:

$$23 \times 9 = (23 \times 10) - 23 = 230 - 23$$
$$= 230 - 20 - 3 = 210 - 3 = 207,$$

$$40 \times 9 = (40 \times 10) - 40 = 400 - 40 = 360,$$

Multiply by 9, 19, 29, ···

$$88 \times 9 = (88 \times 10) - 88 = 880 - 88$$
$$= 880 - 80 - 8 = 800 - 8 = 792\,,$$

$$145 \times 9 = (145 \times 10) - 145$$
$$= 1450 - 145 = 1450 - 100 - 40 - 5$$
$$= 1350 - 40 - 5 = 1310 - 5 = 1305\,.$$

Similarly, to multiply a number by 19 just multiply it by 20 and subtract the number itself, in fact

$$19 = 20 - 1\,.$$

Let's see some examples:

$$23 \times 19 = (23 \times 20) - 23$$
$$= (23 \times 2) \times 10 - 23 = 46 \times 10 - 23$$
$$= 460 - 23 = 460 - 20 - 3 = 440 - 3 = 437\,,$$

Multiply by 9, 19, 29, ⋯

$$56 \times 19 = (56 \times 20) - 56$$
$$= (56 \times 2) \times 10 - 56$$
$$= (50 \times 2 + 6 \times 2) \times 10 - 56$$
$$= (100 + 12) \times 10 - 56 = 1120 - 56$$
$$= 1120 - 20 - 30 - 6 = 1100 - 30 - 6$$
$$= 1070 - 6 = 1064\,,$$

or

$$83 \times 19 = (83 \times 20) - 83$$
$$= (83 \times 2) \times 10 - 83$$
$$= (80 \times 2 + 3 \times 2) \times 10 - 83$$
$$= (160 + 6) \times 10 - 83 = 1660 - 83$$
$$= 1660 + 60 - 20 - 3 = 1600 - 20 - 3$$
$$= 1580 - 3 = 1577\,.$$

The multiplication of a number by 29 (and, similarly, by 39, 49, 59 and so on) can be performed by observing that

$$29 = 30 - 1 = 3 \times 10 - 1,$$

whereby the number is multiplied by 30 and then the number itself is subtracted to the result. For example

$$6 \times 29 = (6 \times 30) - 6 = 180 - 6 = 174,$$

or

$$22 \times 29 = (22 \times 30) - 22$$
$$= (22 \times 3) \times 10 - 22$$
$$= 66 \times 10 - 22 = 660 - 22$$
$$= 600 - 20 - 2 = 580 - 2 = 578.$$

18) Factorization

Another technique of fundamental importance is to factor the various numbers before carrying out any calculation. Suppose we want to calculate the product

$$18 \times 15,$$

let's factor the two numbers, 18 can be written as

$$18 = 2 \times 9 = 2 \times 3 \times 3,$$

while for 16 we have

$$15 = 3 \times 5,$$

hence the product

$$18 \times 15 = 2 \times 3 \times 3 \times 3 \times 5.$$

We observe that we can multiply the 2 and the 5 obtaining 10 to simplify the calculation, in fact

$$18 \times 15 = 3 \times 3 \times 3 \times 10 = 27 \times 10 = 270.$$

Let's see other examples:

$$160 \times 25 = (2 \times 80) \times (5 \times 5)$$
$$= (80 \times 5) \times 10 = 400 \times 10 = 4000,$$

Factorization

$$190 \times 30 = (2 \times 95) \times (6 \times 5)$$
$$= (95 \times 6) \times 10 = (90 \times 6 + 5 \times 6) \times 10$$
$$= (540 + 30) \times 10 = 570 \times 10 = 5700,$$

$$18 \times 500 = (2 \times 9) \times (5 \times 100)$$
$$= (9 \times 100) \times 10 = 900 \times 10 = 9000.$$

19) Square of a number

The square of any number can be obtained by using the square of a binomial, writing the starting number as an appropriate sum of two numbers.

For the square of a sum (square of a binomial) we have

$$(a+b)^2 = (a+b) \times (a+b) = a^2 + b^2 + 2 \times a \times b,$$

which can be also read as "the square of the first added to the square of the second added to the double product of the two numbers".
In our case the idea is to decompose a number as the sum of two numbers for which you

know how to calculate their squares. For example:

$$23^2 = (20+3)^2 = 20^2 + 3^2 + 2 \times 20 \times 3$$
$$= 400 + 9 + 6 \times 2 \times 10$$
$$= 400 + 9 + 120 = 520 + 9 = 529.$$

For this purpose, we report the squares of multiples of 10 in the following table:

10^2	20^2	30^2	40^2	50^2	60^2	70^2	80^2	90^2
100	400	900	1600	2500	3600	4900	6400	8100

Let's focus now on the square of a number ending with the digit 5. For example 115. The trick is to remove the 5 from the number and multiply the remaining number (i.e. 11) by its next (in this case 12) and add to the end (to the right) of the obtained number the digits 2 and 5. For example

$$11 \times 12 = 12 \times 10 + 12 = 120 + 12 = 132,$$
$$115^2 = 115 \times 115 = 13225.$$

Let's clarify with some other examples:

- $35^2 = 35 \times 35 = 1225$ (because $3 \times 4 = 12$, i.e. we remove the 5 from 35, getting 3 and multiply it by its next, 4, and then add "25" at the end);

- $75^2 = 75 \times 75 = 5625$ (because $7 \times 8 = 56$, i.e. we remove the 5 from 75, getting 7 and multiply it by its next, 8, and then add at the end "25");

- $185^2 = 185 \times 185 = 34225$ (because $18 \times 19 = 342$, i.e. we remove the 5 from 185, getting 18 and multiply it by its next, 19, and then add at the end "25").

Using all these rules we can calculate the squares of many numbers, in fact, just as an example, using the results above, we can write:

$$39^2 = 39 \times 39 = (35+4)^2$$
$$= 35^2 + 4^2 + 2 \times 4 \times 35$$
$$= 1225 + 16 + 70 \times 4 = 1225 + 16 + 280$$
$$= 1200 + 200 + 80 + 25 + 16$$
$$= 1480 + 41 = 1521,$$

$$76^2 = 76 \times 76 = (75+1)^2$$
$$= 75^2 + 1^2 + 2 \times 1 \times 75$$
$$= 5625 + 1 + 75 \times 2 = 5625 + 1 + 150$$
$$= 5600 + 150 + 25 + 1 = 5750 + 26 = 5776.$$

20) Units in a product

When we have to decide, for example in a multiple choice test, the correct result of a product, we can also proceed, as a first approach, by exclusion.

For example, the product of the ones digits of two numbers provides information on the one digit of their product. For example in the product

$$8742 \times 328,$$

the product of the ones is

$$2 \times 8 = 16$$

and we can conclude that the result of the starting product has as its ones the number 6, which is the one digit of the product of the ones of the two factors.

We can then answer the initial question, i.e.

- Which of the following answers represents the correct result for the product 526 × 812?

  a) 419374;

  b) 427112;

  c) 430813;

  d) 434365.

the answer is immediate, because making the product of the ones of the two factors 526 and 812 we obtain

$$6 \times 2 = 12$$

and therefore the one of the product must be

Made in the USA
Las Vegas, NV
12 June 2021

24652778R10069